ALIS

2661921

Space

# Stars

## Charlotte Guillain

Heinemann LIBRARY

**www.heinemannlibrary.co.uk**
Visit our website to find out more information about Heinemann Library books.

To order:
☎ Phone 44 (0) 1865 888066
🖹 Send a fax to 44 (0) 1865 314091
💻 Visit the Heinemann Bookshop at www.heinemannlibrary.co.uk to browse our catalogue and order online.

Heinemann Library is an imprint of Capstone Global Library Limited, a company incorporated in England and Wales having its registered office at 7 Pilgrim Street, London, EC4V 6LB – Registered company number: 6695582

Heinemann is a registered trademark of Pearson Education Limited, under licence to Capstone Global Library Limited

Text © Capstone Global Library Limited 2009
First published in hardback in 2009
The moral rights of the proprietor have been asserted.

Edited by Siân Smith, Rebecca Rissman, and Charlotte Guillain
Designed by Joanna Hinton-Malivoire
Picture research by Tracy Cummins and Heather Mauldin
Production by Duncan Gilbert
Originated by Heinemann Library
Printed and bound in China by South China Printing Company Ltd

ISBN 978 0 431 02046 4
13 12 11 10 09
10 9 8 7 6 5 4 3 2 1

**British Library Cataloguing in Publication Data**
Guillain, Charlotte
    Stars. - (Space)
    1. Stars - Juvenile literature
    I. Title
    523.8

**Acknowledgements**
We would like to thank the following for permission to reproduce photographs: Alamy pp. **7** (©B.A.E. Inc.), **14** (©Peter Arnold, Inc.); Getty Images pp. **4** (©Stephen Alvarez ), **5** (©Suk-Heui Park), **6** (©Stocktrek), **8** (©Carlos Emilio); NASA pp. **9** (©SOHO), **13** (©Don Figer, STScI), **18** (©JPL-Caltech/GSFC/SDSS), **20**, **23a** (©JPL-Caltech/STScI), **23b** (©SOHO); Photo Researchers Inc pp. **10** (©Mark Garlick), **11** (©Jean-Charles Cuillandre), **12** (©L. Dodd), **15** (©John Chumack), **16** (©Mark Garlick), **19** (©Chris Butler); Photolibrary pp. **21**, **22**, **23c** (©Nick Dolding); Shutterstock p. **17** (©Mahesh Patil).

Front cover photograph reproduced with permission of NASA (©CXC/JPL-Caltech/CfA ). Back cover photograph reproduced with permission of Shutterstock (©Mahesh Patil).

Every effort has been made to contact copyright holders of material reproduced in this book. Any omissions will be rectified in subsequent printings if notice is given to the publishers.

# Contents

# Space

Stars are in space.

4

Space is up above the sky.

# Stars

Stars are a long way from Earth.

We can see stars shining in the sky.

Stars look very small in the sky.

But stars are huge balls of gas.

Stars are very hot.

Stars shine brightly.

Stars can be different sizes.

Stars can be different colours.

Very big stars are blue or red.

14

Smaller stars are yellow.

15

The Sun is a yellow star.

the Sun

Earth

The Sun is the nearest star to Earth.

# Galaxies

A galaxy is a group of stars.

The Sun is in a galaxy called the
Milky Way.

19

There are many galaxies in space.

telescope

We need a telescope to see
most galaxies.

# Can you remember?

What is this?

22

Answer on p.24

# Picture glossary

**galaxy**  a group of stars. The galaxy we live in is called the Milky Way.

**gas**  not solid like wood or liquid like water. Air is a gas that we breathe in but cannot see.

**telescope**  something you look through which makes far away things look bigger and easier to see

# Index

Answer to question on p.22: A telescope.

**Notes for parents and teachers**
**Before reading**
Ask the children if they have ever looked up into the sky at night and seen the stars. On a clear night you can see hundreds of stars twinkling in the sky. Explain that although stars look small this is because they are very far away. Stars are very big indeed and very hot and that is why they shine brightly.

**After reading**
• Make a magic star picture. Draw stars using a white wax crayon on white paper. Then paint over the stars with a dilute mix of black poster paint and watch the stars appear in the night sky.

• Make a 3D hanging star. Give each child two five point stars made of sturdy white card. Tell them to colour them with yellow crayons. Cut a slit in one star from the tip of one of the points to near the centre of the star and in the other star from the lowest part of a point, across the centre and towards a facing point. Slot the two stars together. Hang the stars from the ceiling.